PROJECT NOTEBOOK

Copyright 2015

All Rights reserved. No part of this book may be reproduced or used in any way or form or by any means whether electronic or mechanical, this means that you cannot record or photocopy any material ideas or tips that are provided in this book.

Printed in the USA
CPSIA information can be obtained
at www.ICGtesting.com
LVHW081625121124
796426LV00014B/780

* 9 7 8 1 6 8 1 4 5 5 8 2 2 *